Freeport Public Library

314 W. Stephenson
Freeport, IL 61032
815-233-3000

Each borrower is held responsible for all materials charged on his
card and for all fines accruing on the same.

**The Kids'
Library of
Personal Safety**

A Kid's Guide
to Staying Safe Around

Maribeth Boelts

The Rosen Publishing Group's
PowerKids Press™
New York

Published in 1997 by The Rosen Publishing Group, Inc.
29 East 21st Street, New York, NY 10010

First Edition

Book Design: Erin McKenna

Photo Credits: p.4 © Eric Berndt/MIDWESTOCK; p.7 © David E. Spaw/MIDWESTOCK; p.8 © Scott Cook/MIDWESTOCK; p.11 © Steven Ferry; p.12 © J. Myers/H. Armstrong Roberts, Inc.; p.15 © Bob Greenspan/MIDWESTOCK; p.16 © S. Feld/H. Armstrong Roberts, Inc.; p.19 © Robert W. Slack/International Stock Photography; p.20 © J. Patton/H. Armstrong Roberts, Inc.

Boelts, Maribeth, 1964–
    A kid's guide to staying safe around fire /by Maribeth Boelts.
        p.   cm. — (The kids' library of personal safety)
    Includes index.
    Summary: Discusses the dangers of a fire, what to do if caught in a fire, and how to plan an emergency exit for a family.
    ISBN 0-8239-5077-8 (lib. bdg.)
    1. Fire prevention—Juvenile literature. 2. Fires—Juvenile literature. [1. Fire prevention. 2. Fires. 3. Safety.] education
    1. Title. 11. Series.
    TH9148.B64 1996
    628.9'2—dc21   96-47918
                                                                                          CIP
                                                                                          AC

Manufactured in the United States of America

Contents

Fire

Fire has a lot of power. It gets its power and makes heat and light by **burning** (BERN-ing). Fire can burn just about anything. We use fire for many things. It helps us to cook our food. Fire can keep us warm. And it can help us to see with its light.

But fire can also be very **dangerous** (DAYN-jer-us). It can burn a person, a person's house, or a whole forest. That's why it is important to learn how to be safe around fire.

◀ Fire has been used by people for thousands of years.

Matches and Lighters

Matches (MACH-ez) and **lighters** (LY-terz) are **tools** (TOOLZ) with an important job. A lighter can light candles on a birthday cake. Matches can light a campfire. But matches and lighters are not toys. They can start fires!

Matches and lighters should only be used by adults. If you see a child playing with them, tell an adult right away. Matches and lighters can cause a harmful fire if they are not used the right way.

Matches can be dangerous if they are not used the right way. ▶

Fire Indoors

We use fire indoors in the kitchen. Your parents cook food in the kitchen. They may use the oven, or cook something in a pan on top of the stove. If you have an electric stove, you won't see fire. If your parents have a gas stove, you will see blue **flames** (FLAYMZ) when you turn it on. It is important never to put anything in that flame. And if your parent is cooking with oil or grease, someone should always watch the pan. Grease gets hot very fast, and can sometimes catch fire. If this happens, tell an adult right away.

You won't see fire on an electric stove, but it still gets very hot.

Fire Outdoors

We use fire outdoors to cook and to keep warm. Your parents may cook hamburgers on your backyard grill. Or if you're camping, your family may build a campfire to stay warm. It's an adult's job to watch the fire. You can help by remembering to play away from grills or campfires.

Another way fire is used outdoors is in fireworks. Fireworks are fun to watch. They can also be dangerous. Many people get hurt and many fires are started by fireworks.

Cooking food on an outdoor grill is fun, as long as everyone remembers to be careful. ▶

Every Fire Is Dangerous

A fire can start from something as simple as a candle tipping over onto a table or from grease getting too hot in a pan on the stove. And a fire can spread very quickly. Your whole house can be on fire in just a few minutes. Fires can get as hot as 600° F. Smoke from a fire is dark. It is very hard to see through smoke. And it is hard to breathe air that has smoke in it. Every fire is dangerous, so fire must be handled carefully.

◀ Even small fires can cause big problems. That's why it's important to be careful around all kinds of flames.

13

Where There's Smoke, There's Fire

If you see smoke, there is probably a fire nearby. Smoke is dangerous. It can contain **poisonous** (POY-zun-us) gases that make it hard to breathe. During a fire, more people are hurt from the smoke than from flames.

If you need to get out of a fire through smoke, you should *crawl on your hands and knees under the smoke*. The air is clearer, cleaner, and cooler underneath. But do not crawl on your belly. Some of the poisonous gases may be in a thin layer on the floor.

14

It's a fireman's job to put out a fire ▶ and to make sure no one is hurt.

The Great Escape

E.D.I.T.H. means Exit Drills In The Home. Every family should have a plan that tells how to get out of the house in case there is a fire. Think about these things when you and your family make your E.D.I.T.H. plan:

- Are there two ways to get out of every room? Remember that one way may be a window if the room is close to the ground.
- Where will everyone meet outside?
- Who will call 911, and from where?

When you finish the plan, your family should practice it. Practice makes perfect!

◀ Smoke detectors are another way to protect you and your family from a fire.

17

Nighttime Fire Safety

There are some things you can do to stay safe from a fire that may happen at night.

- If you can, sleep with your bedroom door closed.
- If there is a fire, roll out of bed and crawl low under the smoke. Yell "Fire!" as loud as you can.
- Crawl to the door and feel the door and the doorknob with the back of your hand. Is it hot? *If it is, do not open the door.* Block the bottom of the door with a blanket. Crawl to another exit.

There are steps you can take to keep yourself safe if there is a fire at night. ▶

Your Safety is First

Now you know what to do during a fire if the bedroom door is hot. But what if the door is cool? You should open it just a little bit. Be ready to close the door fast if there is heat and smoke. Look into the hall. If it is clear, crawl out of your bedroom. Take short breaths so you don't choke on any **fumes** (FYOOMS).

Remember: Never go back into a room to get something. Your job is to get safely out of the fire and stay out!

◀ If there is a fire, the only thing to worry about getting out of your room or house is you!

How Do I Call for Help?

If there is smoke or fire at your house, you need to **report** (ree-PORT) it right away. Call 911 from your neighbor's telephone. Stay calm and speak slowly. Tell them your name and the address of the fire.

Stay on the telephone until the person on the other end tells you to hang up.

Then go back to your family meeting place and wait for help.

Fire can be scary. But if you're careful, and if you remember the rules of fire safety, fire can also be useful.

22

Glossary

burning (BERN-ing) To use up fuel and give off heat, light, and gases.

dangerous (DAYN-jer-us) Something that is not safe.

flame (FLAYM) A glowing, hot part of a fire.

fumes (FYOOMS) Gas or smoke that is harmful.

lighter (LY-ter) A tool used to light something with fire.

match (MACH) A short piece of wood with a tip that catches fire when scraped on a rough surface.

poisonous (POY-zun-us) Something that is harmful and can cause sickness.

report (ree-PORT) To tell someone.

tool (TOOL) Something that is used for work.

23

Index